Christmastide

Twelve carols for mixed voices

G000134660

Contents

Where first lines differ from titles the former are shown in italics.

Carols suitable for unaccompanied singing are marked thus *.
Carols with instrumental material available on hire are marked thus †.

Oxford University Press Music Department, Walton Street, Oxford OX2 6DP

To the nuns of St. Mary's Convent, York

A MAIDEN MOST GENTLE

Words paraphrased from
The Venerable Bede
by Andrew Carter

French tune
arranged by
ANDREW CARTER

1. A maid-en most gen-tle and

ten-der we _ sing: Of Ma-ry the mo-ther of Je-sus our King. A-ve, a-

-ve, a-ve Ma-ri-a, A-ve, a-ve, a-ve Ma-ri-a.

Also available separately (X 266)

Printed in Great Britain

4

-ve Ma - ri - a, A - ve, a - ve, a - ve Ma - ri - a.

Voices unaccompanied

T.

4. Three Kings came to wor-ship with gifts rich and_ rare, And mar-velled in

B. 1
B. 2

awe at the babe in her_ care.

Sopranos

A - ve, a - ve, a - ve Ma-ri-

Altos 1 & 2

A - ve, a - ve Ma-ri-

-a, A - ve, a - ve, a - ve Ma-ri - a.

-a,_ A - ve, a - ve Ma-ri - a.

Ped.

All Voices in Unison *f solo*

5. Re - joice and be glad at this Christ - mas we_

For Mary

ADAM LAY YBOUNDEN

Words anon. 15th century

PHILIP LEDGER

Also available separately (X 276)

CHRISTMAS TIME

Words and music by
PENELOPE HUGHES
arranged by
DAVID WILLCOCKS

(repeat with octaves ad lib.)

Also available separately (X 277)

This carol won the first prize in The Bach Choir 1980 Carol Competition for children.

The accompaniment is scored for brass a 8 and percussion: material is on hire.

Penelope Hughes was born in 1971.

(Piano may double semi-chorus)

p **TUTTI SOPRANOS**
Ah

mp **TUTTI ALTOS**
Al – le – lu – ia,

(8)
Child was born to take our sins a – way?
Lord and Christ and Sa – viour of us all.

p

S.
Ah _____ Ah

A.
al – le – lu – ia, al – le – lu – ia. Ah _____

mf

T.
f
Al – le – lu – ia,

B.
f
Al – le – lu – ia,

mf

FALAN-TIDING

Words *c.* 1610

Tyrolese carol
arranged by
JOHN R. WOOD

Also available separately (X 285)

in a sil - ly man - ger poor, Be - twixt an ox and ass, Whom

these three— kings did all a - dore As God's high plea - sure was.

attacca v. 3

S.
A.

3. And

T.
B.

ORGAN

Gt.

Sw.

Ped.

for the joy of his great birth A thou - sand an - gels— sing: 'Glo -

Gt.

For JoAnne

MARY'S LULLABY

Words and music by
JOHN RUTTER

This carol is scored for flute, oboe, harp, and strings. Scores and parts are on hire.

It has been recorded by the Choir of Clare College, Cambridge, on Argo ZRG 914.

Also available separately (X 272)

© Oxford University Press 1979

NIGHTINGALE CAROL

Words by
MARY GAIR

Melody by
WILHELM DÖRFLER
arranged by
ANDREW CARTER

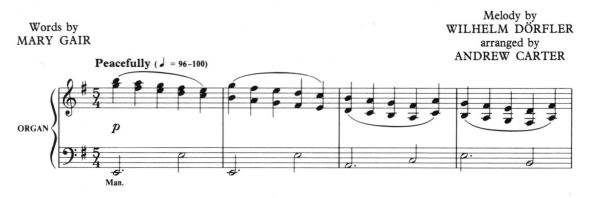

1. Night - in - gales in wood - land call - ing Sum - mer songs though snow is fall - ing;

Je - sus Christ my heart is warm - ing, Born on win - ter's sum - mer

Melody used by kind permission of Otfried Doerfler

Also available separately (X 282)

22

ONCE IN ROYAL DAVID'S CITY

Words by
C. F. ALEXANDER

H. J. GAUNTLETT
harmonized by A. H. MANN (vv. 1–5)
descant and organ part by
PHILIP LEDGER

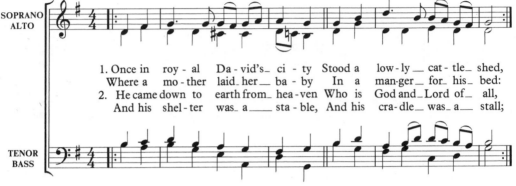

1. Once in roy - al Da - vid's_ ci - ty Stood a low - ly _ cat - tle - shed,
Where a mo - ther laid_ her _ ba - by In a man-ger_ for_ his_ bed:
2. He came down to earth from _ hea - ven Who is God and_ Lord of_ all,
And his shel - ter was_ a _____ sta - ble, And his cra - dle_ was_ a _ stall;

Ma - ry_ was that mo - ther mild, Je - sus_ Christ_ her lit - tle_ child._
With_the_ poor and mean and low - ly_ Lived on_ earth_ our Sa - viour ho - ly.

3. And through all his wondrous childhood
   He would honour and obey,
   Love and watch the lowly maiden,
   In whose gentle arms he lay:
   Christian children all must be
   Mild, obedient, good as he.

4. For he is our childhood's pattern,
   Day by day like us he grew,
   He was little, weak, and helpless,
   Tears and smiles like us he knew:
   And he feeleth for our sadness,
   And he shareth in our gladness.

5. And our eyes at last shall see him,
   Through his own redeeming love,
   For that child so dear and gentle
   Is our Lord in heaven above;
   And he leads his children on
   To the place where he is gone.

DESCANT

6. Not in that poor low - ly sta - ble, With the
We shall see him; but in hea - ven, Set at

OTHER VOICES

ORGAN

ox - en stand - ing by,
God's right hand on high;
Where like stars his chil - dren

crowned All in white shall wait a - round.

STILL, STILL, STILL

English translation by
MEG PEACOCKE

German carol
arranged by
PHILIP LEDGER

Also available separately (X 284)

rall.

THREE KINGS OF ORIENT

Words and Music by
J. H. HOPKINS
arranged by JOHN RUTTER

Stately, like a royal procession (♩. = 50)

*Instrumentation: 2 Fl, 2 Ob, 2 Cl, 2 Bsn, 2 Hn, Perc, Hp, Strings. Orchestral scores and parts are on hire.

Also available separately (X 283)

© Oxford University Press 1981

(Hum) are, Bear-ing gifts we tra-verse a - far, Field and

(Hum) foun - tain, Moor and moun - tain, Fol - low -ing yon - der star:

A Refrain for Verses 1, 2 and 3

O _____ star of won - der, star of night, Star with

royal beauty bright; Westward leading, Still pro-

-ceeding, Guide us to thy perfect light.

B Verses 2 and 3

V. 2: TENOR SOLO or SEMI-CHORUS
V. 3: BARITONE SOLO or SEMI-CHORUS

2. Born a king on Beth-le-hem plain, Gold I bring to crown him a - gain,
3. Frank-in-cense to of-fer have I, In-cense owns a De-i-ty nigh,

p 2. Hum
mp 3. Aw

King for ev - er, Ceas - ing nev - er, O - ver us all to reign:

Prayer and prais - ing, All men rais - ing, Wor-ship him, God most high:

C *Verse 4*

BASS SOLO or SEMI-CHORUS

4. Myrrh is mine; its bit - ter per - fume

Breathes a life of ga - ther - ing gloom; Sor - row - ing,

ritard. **a tempo**

sigh - ing, Bleed - ing, dy - ing, Sealed in the stone - cold tomb:

O _____ star of won - der, star of night, Star with

(Accompt. tacet)

roy - al beau - ty bright; West - ward lead - ing,

Still pro - ceed - ing, Guide __ us to __ thy per - fect light.

5. Glo - rious now be -

-hold him a - rise,_____ King, and God,_____ and sa - cri-

-fice!_____ Heav'n sings Al — le - lu — ia: Al — le-

-lu — ia the earth__ re - plies: O_____ star of

To Lynne Dawson

SPANISH CAROL

English text by
ANDREW CARTER

Traditional
Arranged by
ANDREW CARTER

Also available separately (X 281)

For the choir of Clare College, Cambridge

THE HOLLY AND THE IVY

English traditional carol
arranged by
JOHN RUTTER

Also available separately (X 271)

This arrangement is scored for flute, oboe, harp, and strings (2, 2, 2, 1, 1 or more).

Scores and parts are on hire.

Recorded by the Choir of Clare College, Cambridge on Argo ZRG 914

hol - ly bears the crown. _____ O the ris - ing of the sun ___ And the run - ning of the deer, _____ The _ play - ing of the mer - ry or - gan, Sweet sing - ing in the choir. _____

2. The hol - ly bears a blos - som As white as a - ny flow'r; _____

And_ Ma-ry bore sweet Je-sus Christ To_ be our sweet Sa - viour.

O the ris-ing of the sun_ And the run-ning of the deer,_

The_ play-ing of the mer-ry or-gan, Sweet sing-ing in the choir._

legato

WATTS'S CRADLE SONG

ISAAC WATTS
(1674–1748)

BRYAN KELLY

★Scores and string parts are on hire.

Also available separately (X 264)

VERSE
Più mosso

SOPRANO
ALTO

1. How much bet - ter thou art at -tend- ed Than the Son of God could be
2. Soft and ea - sy is thy cra - dle; Coarse and hard thy Sa - viour lay,
3. May'st thou live to know and fear him, Trust and love him all thy days:

TENOR
BASS

dim. e rit. _ _ _ _ _ _ *p*

When from hea - ven he de - scen- ded And be - came a child like thee!
When his birth-place was a sta - ble And his soft - est bed was hay.
Then go dwell for ev - er near him, See his face and sing his praise!

dim. e rit. _ _ _ _ _ *p*
Repeat CHORUS

CODA

dim. e rit.

(head.)

dim. e rit.

Processed and printed by
Halstan & Co. Ltd., Amersham, Bucks., England